CAREERS IN
GREEN ENERGY

SOLAR AND WIND POWER JOBS

FINDING A NEW JOB, ESPECIALLY IF IT IS a career change, can be tough. One way to increase your odds is to look for opportunities in green energy. It is one of the fastest growing, most innovative sectors of the economy. After years of hype and false starts, the shift to green power has begun to accelerate at a pace that is far beyond the predictions of even the most experienced experts. Wind farms with their huge towers dotting rangelands and hillsides are no longer a novel sight. Solar panels are covering more than homes. They are now blanketing everything from parking lots to airports. The increased usage has driven down the price of energy dramatically, making it competitive with fossil fuels and even natural gas.

The green energy industry has been boosted by big subsidies on wind and solar power and general backing by the federal government since the turn of the 21st century. While federal subsidies are being pulled back, states and local governments are stepping up to push the industry forward. For example, Illinois has new requirements and incentives that are expected to produce dozens of solar farms to feed the state's electric grids. In California, all new homes must be built with solar panels starting in the year 2020.

The shift to cleaner power is disrupting entire industries, which means many of them are scrambling to hire like crazy. Consider this real job opening posted online recently:

Need someone with a flexible schedule who can work out of town. Company provides transportation and fuel expenses for out of town jobs. Little or no experience is ok if willing to learn.
Job Type: Full-time
Salary: $150.00 to $200.00 per day

By far, the most opportunities are for solar panel installers and wind turbine technicians (known as wind-techs). Both are expected to experience rapid growth for the foreseeable future. In fact, solar panel installers and windtechs are the only occupations that are expected to double in size within the coming decade. There are many more solar installers than windtechs, but according to current government data, the fastest-growing job title in the country is "wind turbine technician."

Taking advantage of these opportunities is relatively easy.

There are no educational requirements to get started, and on-the-job training is standard. Solar installers can start in a job out of high school, and so can windtechs – if they have some mechanical skills and a good understanding of how a wind turbine functions. Most windtechs go to technical school for a year or two to learn about wind energy technology. Some choose to complete certificate programs or earn associate degrees in wind turbine maintenance.

Considering the minimal entry requirements, the pay is good. Beginners typically start out earning between $35,000 and $40,000 a year. After some training, that quickly rises to an average of $50,000. With several years of experience, in the right location and with the right kind of employer, it is possible to earn $80,000 or more.

If you are physically fit, have a knack for working with your hands, are not afraid of heights or bad weather, green energy could be the answer to your career choice. In addition to a future you can depend on, you will be rewarded with knowing you are contributing to a cleaner environment and helping reduce the dangers of climate change.

WHAT YOU CAN DO NOW

YOU DO NOT NEED A COLLEGE DEGREE to enter the green energy field, but you do need to graduate from high school or earn an equivalent. Make good use of your time in high school by selecting courses that will help prepare for your future career. Consider taking classes in computer applications, business (many in this field are self-employed), a foreign language like Spanish, physical education, and any classes that will help improve your communications skills.

Learn the basics by taking classes focused on green energy issues and industries at your local community college or technical school. There are also free or inexpensive online classes if you cannot find something local.

A good way to determine if this is a good career for you is to job shadow for a day (or more). Simply call any company in the business and ask if one of their installers or techs would be willing to let you come along with them to work. Sweeten the deal by offering to help out by carrying tools or cleaning up. Maybe offer to buy them lunch in exchange for the opportunity to ask questions and gather advice.

Due to the great demand for new employees in the green energy field, it is possible to start without any experience, but work experience is highly valued by employers. Try to get a part-time job after school or during the summer. If you cannot find a paying job, volunteer work will also provide experience and it looks just as good on a résumé. Start your search for a volunteer position by contacting community conservation and environmental action groups.

Read up on the green energy industry. Learn the common terminology and the basics of how solar and wind power work. Start getting into shape because the work is as physical as it is technical.

HISTORY OF THE CAREER

THROUGHOUT MOST OF HUMAN HISTORY, energy needs were satisfied with natural and renewable resources. Heat came from the sun or by burning plant materials. Food was cooked over open fires. Horses and the wind in sails provided transportation. Animals performed heavy labor. Water and wind drove simple machines that pumped water and ground grain into flour.

Wind was especially important to developing civilizations. Windmills first appeared in the 10th century in the gusty Seistan region of Persia. The first mills served multiple purposes, including grinding corn and irrigating gardens. Their use soon spread to India and China where farmers depended on them for the energy to pump water, grind grains, and crush sugarcane. In the late 16th century, the windmill had been developed to the highest possible degree by Dutch engineers. Now, energy generated by the huge and efficient windmills could do everything from grind fine spices to saw trees into lumber. In the 1800s, windmills were vital to the settling of the West. In many regions, there was more wind than water. The windmill was able to liberate groundwater that was necessary to settle rangeland and produce crops. At one point, there were more than six million windmills dotting the US landscape.

Then came the Industrial Revolution and the discovery of coal. The Industrial Revolution was a boon to civilization and marked the beginning of modern times. It was also a curse. It took massive amounts of energy to power the factories. While mass producing products at a rate never before seen, those same factories spewed out pollution that dirtied the air and the water. It was coal that first fueled the Industrial Revolution in the Western world, until oil was introduced. In 1859, the first successful oil well was drilled in Titusville, Pennsylvania. Few had any idea that petroleum would change the world. The availability of greater quantities of fuel led to an acceleration of technologies well into the 20th century.

Some forward thinkers, however, began to worry that fossil fuels would run out. One of those people was French inventor and physics professor, Augustin Mouchot, who developed a solar powered steam engine that could drive industrial machinery. It was Mouchot's belief that the sun's heat could replace the burning of coal to run Europe's industries. Clearly, he was a man ahead of his time when he wrote: "The time will arrive when the industry of Europe will cease to find those natural resources, so necessary for it. Petroleum springs and coal mines are not inexhaustible but are rapidly diminishing in many places. Will man, then, return to the power of water and wind? Or will he emigrate where the most powerful source of heat sends its rays to all? History will show what will come."

About the same time that Mouchot's invention was making its debut at the 1878 Paris Expo, another solar device was being introduced. Professor William Grylls Adams discovered that an electrical current could be started in selenium solely by exposing it to sunlight. This led to the development of the first selenium solar cell. It was successfully demonstrated, but failed to convert enough sunlight to power electrical equipment. Its significance, however, did not go unnoticed. Well known scientist in the field of electricity, Werner von Siemens, called the discovery scientifically of the most far-reaching importance. He said, "Its practical value will be no less obvious when we reflect that the supply of solar energy is both without limit and without cost, and that it will continue to pour down upon us for countless ages after all the coal deposits of the earth have been exhausted and forgotten."

In the early 20th century, Albert Einstein published the first theoretical work describing the photovoltaic effect. In the paper, Einstein described how light contained packets of energy that could be harnessed and transformed into power. Combined with the discovery of the electron, there was now a better understanding of photo electricity. In 1911, a Scientific American article submitted that "in the far distant future, natural fuels having been exhausted, solar power will remain as the only means of existence of the human race."

Scientists continued to work on developing solar engines until the start of World War I. Others were hard at work, developing machines that could transform wind into electricity. In 1927, the first commercial wind turbines were sold to remote farms around the country. The wind-electric systems were popular, and by the 1940s, there were hundreds of thousands of them in operation.

The 1950s saw advancements in renewable energy technologies as there was increasing warnings of too much dependence on fossil fuels. The first silicon solar cell capable of generating a measurable electric current was developed by Bell Laboratories in 1953. The New York Times touted it as "the beginning of a new era, leading eventually to the realization of harnessing the almost limitless energy of the sun for the uses of civilization." At the same time, the concept of "peak oil" began a new drive towards renewable energy sources. Environmentalists and industrialists alike were alarmed at the exponential growth in human population and accompanying oil consumption. Many argued that renewable energy was not just a scientific innovation for the future, but a necessity. The peak oil debate raged on for years. It was not about whether fossil fuels were inexhaustible, but rather when we would reach the point of demand outstripping supply. New pockets got fewer and smaller, and around 2008, we did indeed reach peak oil.

Until the early 1970s, solar cell technology was too expensive for practical use. By using a poorer grade of silicon and packaging the cells with cheaper materials, the price dropped significantly – from $100 a watt to $20 per watt. Solar cells could now compete with traditionally generated electricity. They were especially welcome in situations where there were no power lines nearby.

From the 1970s through the 1980s, the federal government collaborated with industry to advance green energy technologies. The US Department of Energy established the Solar Energy Research Institute, the first government facility dedicated to harnessing power from the sun, in 1977. It was later renamed the National Renewable Energy Laboratory. The Department of Energy, together with NASA, also worked on developing large commercial wind turbines that could produce multi--megawatt power. The world's first wind farm was installed in New Hampshire in 1980. A year later, the first solar-thermal power plant began operation in California.

The importance of replacing fossil fuels with green energy sources was well established by the beginning of the 21st century. Millions of solar panels were installed on rooftops across the country. Wind turbines sprouted like corn in fields. The more these devices were used, the lower the cost, and the more popular they became. Still, it took a prod from government leaders to really get green energy into a competitive position. In 2014, President Obama announced more than 300 private and public sector commitments to create jobs and cut carbon pollution by advancing solar deployment and energy efficiency. In 2018, under Governor Brown's leadership, California became the first state to require all new homes to have solar power by 2020.

Green energy is expected to be cheaper than fossil fuels by 2020. It is also closing in on natural gas much faster than anticipated. In 2017 alone, solar and wind power accounted for nearly 95 percent of new energy generation in the US, while creating jobs at a rate 12 times faster than the rest of the U.S. economy.

WHERE YOU WILL WORK

NEARLY 250,000 AMERICANS WORK in solar. About 135,000 of them work as solar installers, more than all other types of solar jobs combined. There are more than 9,000 companies employing solar installers. Although there are higher concentrations in states that have the sunniest days like Colorado, Florida, and California, they are located in every US state.

The largest group of employers is solar panel installation companies. These are often plumbing, heating, and air conditioning service firms that have contracted with manufacturers to handle installation and maintenance of their solar products. These companies employ about 43 percent of solar installers. About 20 percent work for electrical contractors and other kinds of contractors handling wiring installations. The solar industry is very open to entrepreneurs and as a result, 20 percent of solar installers are self-employed. These are typically experienced installers who have worked with a variety of solar power systems and now work directly for the property owners or through a large project development firm. The rest are employed directly by manufacturers, by building contractors, and by utility companies.

Because photovoltaic (PV) panels convert sunlight into electricity, most PV installation is done outdoors.

Residential installers work on rooftops and in attics and crawl spaces to connect panels to the electric grid. PV installers who build solar farms work at ground level to build structures to hold the PV panel framework.

Wind Turbine Technicians

Windtechs are in the second largest occupation, behind solar installers, among green energy jobs. There are major wind power installations across more than 40 states, which have created many new jobs for wind turbine technicians, called windtechs. The highest numbers of jobs for technicians are in Texas, California, Iowa, Kansas, and Oklahoma (in that order). Only a handful of Southern states have no employers at the moment, but that is expected to change in the near future.

Unlike solar power, which is usually generated for individual buildings, wind power is generated by huge wind farms. The largest employers of windtechs are therefore electric power generation companies. About one third of all windtechs work for these companies. There is a growing number of service firms that provide contracted repair and maintenance services. These firms employ 23 percent of technicians. Like the solar industry, entrepreneurship is encouraged in the wind energy business. About 17 percent of workers are self-employed. Another 15 percent work in utility system construction. The rest work for various professional, scientific, and technical services related to the wind power industry.

Windtechs usually work outdoors, often at great heights and with a partner. They often need to climb ladders that may be more than 250 feet tall and/or slide down a rope to reach the section of a blade that needs servicing. Other times they work in the confined space of the turbine casing. Windtechs often travel to rural areas where wind farms are typically located.

The majority of windtechs work full time according to

regular maintenance schedules determined by a turbine's hours in operation or by the manufacturer. However, they may be on call to handle emergencies that can arise at any time. Turbines are electronically monitored from a central office 24 hours a day. When a problem is detected, windtechs need to get to the worksite and make the necessary repairs as quickly as possible.

THE WORK YOU WILL DO

Solar Installer

Solar installers, sometimes known as solar technicians are the nucleus of the solar industry. Without their work, the industry would grind to a halt. They are the pros that set up and maintain the complex equipment and wiring that connect a solar energy system to the electrical grid. Most solar installers work on residences and businesses, installing rooftop systems. There is a basic sequence of actions that starts with determining the customer's needs, then giving estimates, reading blueprints, gathering materials, installing solar panels, and maintaining or repairing solar panels.

The size of the company will determine how specialized any solar installer's job is within the overall installation process. Large companies may have sales reps do the initial assessment and write up estimates, while others either perform the installation or repairs. Most employers, however, are small and have their solar installers handle all aspects of the job.

Finding out what a customer needs is the first step in each job assignment. This involves visiting the property and talking to the customer, then figuring out how many and what size solar panels are needed. Most people also have questions because they are unfamiliar with the technical aspects of solar photovoltaic systems. After all questions are answered, a rough estimate is written that generally includes the cost of materials and labor. The customer and solar installer must agree on pricing before a job can begin. A contract is usually signed that spells out the details of that agreement.

Next, plans are drawn that illustrate the system's layout.

Based on the blueprints, the solar installer will assemble the necessary solar panels, tools, and any other essentials to complete the job. For smaller residential rooftops, this is usually a simple step that takes little time. Larger commercial facilities will likely require much more preparation.

In most cases, permits will be needed before proceeding with installation. Permits may be required by the building department, fire department, and sometimes HOAs or conservation groups. The electric company will also need to approve of the plans before allowing connection to the grid.

Now it is time for the most important part – installation. Following the blueprints, the installer will use a variety of hand and power tools to measure, cut, assemble, and bolt structural framing and solar modules. They use common tools like drills, saws, screwdrivers, and wrenches, to attach the panels to roofs and connect the wiring. After the system is in place, it must be tested to make sure it works properly. The solar installer will complete the job by giving the customer instructions on how to operate the system.

Most rooftop solar systems generate more energy than is used in the particular structure. Any excess power is routed into the utility grid so that other nearby utility customers can use it. Depending on state and local laws, electricians may need to make the connection to the grid, and inspect for proper wiring, polarity, and grounding.

For larger installations, solar towers may be used to maximize the amount of generated power. Solar towers are used on solar farms and other commercial properties that have sufficient ground area. Multiple solar panels can be mounted on one solar tower, which adds to the amount of power that can be produced. The towers are motorized to move with the sun, which adds an additional boost in electrical output. Working on a solar tower is basically the same as working on a rooftop system – it is just a much bigger and more complex job.

Most solar systems require minimal maintenance, but anything that is exposed to the elements for long periods of time can eventually malfunction. Solar installation companies generally provide maintenance agreements to customers that stipulate repairs will be done as needed. In most cases, it is an installer who performs the repairs. This usually means troubleshooting the problem and replacing old panels or fixing a wiring issue.

Wind Turbine Technicians

Wind turbines are large, sophisticated machines that convert wind energy into electricity. The three major components of a turbine are the tower, three blades, and the nacelle. The nacelle is the part that windtechs are most concerned with. It consists of an outer case, generator, gearbox, and brakes. Windtechs are responsible for installing, maintaining, and repairing all three components. Of these tasks, maintenance is the most important. If a part fails, the wind turbine has to be shut down until it is fixed. Since that costs the owner money, it is better to prevent problems before they occur.

Maintenance involves regularly scheduled equipment inspections, sensor calibration, cleaning, and replacement of aging or malfunctioning components. Much of the daily maintenance work is done in the nacelle, which is very compact. With little room to work inside the nacelle, windtechs will clean and lubricate the shafts, bearings, gears, and other parts. They use handheld power tools and electrical measuring instruments to examine the generator for any faults.

Wind turbines are huge, sometimes more than 250 feet tall. Windtechs use safety harnesses to climb the towers. On any given day, they may climb and inspect a dozen turbines. In some cases, they have to carry new parts while climbing to where they will be installed. While most of the work is performed inside the cramped space of the nacelle, sometimes windtechs must work outside, on top of it. For example, they may need to replace the instruments that measure wind speed and direction. That means the windtech is outside, hundreds of feet off the ground, exposed to winds and other elements. To work safely, they attach their harnesses to rings on the nacelle and carefully follow safety procedures.

Routine maintenance is usually scheduled for three times a year. In between, nacelles are often electronically monitored from the ground where windtechs keep watch in a central office 24 hours a day.

STORIES OF GREEN ENERGY PROS

I Work on a Wind Farm

"I was in technical school when I learned about an opportunity at a career fair for a technician position at a wind farm in west Texas. I applied, knowing that getting the job would mean moving to a sparsely populated area or accepting a very long commute. I had driven past wind farms, but when I saw this one up close it made a big impression. There are 350 turbines producing enough electricity for 45,000 homes. It takes a dozen windtechs to take care of them.

I performed scheduled maintenance and inspections for about three years before joining the troubleshooting team. Now instead of routine, every day is a little different. I spend my days figuring out why a turbine isn't running right. That can be challenging, but once I solve the problem it's very satisfying. In my job, solving those puzzles outdoors in the countryside, working with my hands is very enjoyable.

I love my job and highly recommend it to anyone who has a serious interest in green energy technology. But it's not for everyone. It takes a special kind of person to service multiple turbines in a day, high in the air while repairing blades, and dealing with extreme temperatures and long hours. I think of us as industrial athletes that are in peak physical condition and smart enough to solve the technical challenges that come our way. If that sounds like you, my advice is to give it a try. There are a lot of internship programs that will give you a good sense of what your career will really be like. Now is a good time to start. The wind industry is changing and wind power is here to stay and growing every year."

I Am a Full Service Solar Technician

"My interest in solar energy began in high school when my family had a solar system installed on the roof of our suburban home. My interest grew as I started tinkering around with little solar projects like powering my cell phone and radio. When I graduated, I started working for a large construction company known for their 'green buildings.' I was surprised how easy it was to land the job as an installer. Later, my boss told me that it was a combination of my knowledge of how solar energy worked and my enthusiasm that won him over.

I started as an installer's helper, learning the installation process step by step. After about six months, I was promoted to full-fledged installer. I enjoyed the work and made good money considering I was only 19 years old. Then one day I saw a notice in the office about an opening for a salesperson. The company usually kept the jobs of sales and installation separate, but I convinced my boss that I could do both. It was the best move I could make. I nearly doubled my income and got the best of both worlds - working outdoors with my hands and sharing my passion for solar energy with new customers.

This job is a great fit for me. I have job security and good pay while providing clean, efficient energy that will help reduce pollution and fight climate change."

PERSONAL QUALIFICATIONS

THE KNOWLEDGE AND SKILLS NEEDED to do this kind of work are easily obtained. Employers usually have some kind of on-the-job training available, either formal or informal. Mechanical aptitude is needed to pick up the skills quickly. Windtechs must understand and be comfortable with maintaining and repairing all of the mechanical, hydraulic, braking, and electrical systems of a turbine. Solar installers also work with complex mechanical equipment and electrical systems. Those with a knack for working with tools and mechanical things will find the work easy to learn and enjoyable.

Knowledge is a necessity, but this is physical work. Wind techs and solar installers often lift heavy equipment, parts, materials, and tools in excess of 50 pounds. They also climb up and down ladders many times throughout the day. At the very least they are climbing up to rooftops, but some towers are more than 260 feet high. It takes a lot of stamina to do that, especially while carrying a load. Sore backs and shoulders are common. Being physically fit is essential to avoid serious injury from the strains of everyday work.

Paying attention to details is important. Windtechs are responsible for recording all of the services they perform. Maintenance records need to be complete and accurate, and include precise measurements and every operational step taken. Solar installers have to follow manufacturer instructions with precision. Failing to do so will make the system inoperable. It is also important to follow safety procedures to the letter, never skipping over a single step.

ATTRACTIVE FEATURES

NEW CAREERISTS ARE MOST ATTRACTED to green energy jobs because that is where the fastest job growth is. The world is undergoing a global energy transition from fossil fuels to renewable sources. This transformation in the energy sectors has deep implications in labor markets, both in the US and elsewhere all over the world. It is hard to imagine an occupation that is doubling in size in a single decade. Thousands of new green energy jobs are created every day. Employers are reporting great difficulty filling all the new jobs that are being created.

Taking advantage of the abundance of opportunities is relatively easy. You do not need a college degree. Solar installation jobs require no education beyond high school – employers typically train workers on the job. There are no educational requirements for windtechs either, though one or two years of training are usually needed, depending on their background. There are very few careers that offer such good pay to people with as little education.

Green energy workers have the satisfaction of knowing they contribute to a better life for all. They help save the environment by creating clean energy, reducing the risk of global climate change, and eliminating pollution from carbon emissions. Plus, their work helps stimulate the local economy and creates more indirect jobs. After all, solar panels do not manufacture themselves. Wind turbines do not erect themselves. These activities require intensive labor. Plus, green energy plants and farms generate a new demand for goods and services in the areas where they are located, stimulating local economy and creating indirect jobs. All in all, the green energy sector minimizes threats to the environment, provides more jobs, and promotes a healthy, happy population.

UNATTRACTIVE ASPECTS

THIS IS PHYSICAL WORK THAT brings with it certain risks. Solar installers can fall from ladders and roofs, sustain electrical shocks or burns from hot equipment and materials. Windtechs work at even greater heights and carry even heavier loads of tools. Sometimes they have to rappel down a rope to get to the section of a blade that needs to be worked on. Both of these workers use fall protection equipment and/or wear harnesses.

Just about anyone with a little mechanical know-how and

a desire to do so can get a job as a solar installer or windtech. There is little education required. However, additional training may be needed to advance into the best jobs with the highest pay.

Green energy sources are renewable, clean, and if used more widely could reduce the threat of global climate change. But not everyone is fully convinced of the value of these sources and changing their minds can be an uphill battle. Some believe that the cost of initial setup is too high and that traditional fossil fuels still offer the biggest bang for the buck. Today, solar and wind are the cheapest energy sources around. Others do not believe climate change is real and consider it a waste of resources to address it. There is also some cultural and political pushback from people who want to stick with traditional industries and jobs. Green energy got its big boost from federal subsidies and some of that is being walked back. But the industry has reached critical mass and many states have stepped in with helpful support. Disruption is not easy, and global acceptance of green energy on a mass scale will take time.

The number of jobs is increasing fast, but jobs are not dispersed evenly throughout the country. The introduction of solar and wind energy generation is subject to atmospheric conditions and some geographical locations are better than others. That is steadily changing as the related technologies become more efficient. In the meantime, it may be necessary to relocate in order to find a job.

EDUCATION AND TRAINING

THERE ARE MULTIPLE PATHS TO BECOMING A **solar installer**. Most do not involve college and there are no specific education requirements. Some employers expect to see a high school diploma or GED, but some will start training high school students before they graduate. Some prefer to provide their own formal training on site, while others are not set up for that. They may only hire someone who has already had some training prior to working. In some cases, new employees are referred to solar system manufacturers who provide training on the proper installation methods for the specific products they make.

Most workers train while working on-the-job under the supervision of more experienced employees. Depending on the program and the job duties, training can last anywhere from one month to one year. During that time, beginning solar installers will learn about tools, solar system installation techniques, and safety procedures. Experience in some type of construction may shorten the training time. Electricians are highly valued since the work is closely related. Roofers, carpenters, or even construction laborers have enough knowledge and skills to perform basic duties with minimal direction. In some cases, union electricians and roofers can get access to training modules through apprenticeships. Non-union candidates with prior construction experience can also get up to speed by taking online training courses.

Another option is to take courses at a community college or trade school. Solar panel installation courses cover everything from basic safety to solar system design. Depending on the depth of knowledge desired, a course can last from a few weeks to several months. Generally, the longer the training, the more attractive the job candidate is to prospective employers.

Military veterans have an opportunity to get free training from the Solar Ready Vets program. This program is a joint effort of the US Departments of Defense and Energy to provide the necessary training and connect veterans with jobs in the solar industry. It is designed to meet the needs of high-growth solar employers while building on the technical skills that veterans acquired through their service. Classes last between four and six weeks in a classroom setting.

No license is required other than a clean driver's license.

Windtechs

There is no one way to be trained as a windtech, and there are no specific education requirements. However, most employers expect new windtechs to have mechanical skills and a solid understanding of how a turbine functions. Sometimes candidates who have worked as technicians in other industries can meet those minimal standards and can be trained on-the-job if the employer is willing. Most windtechs learn their trade by attending a technical school before being hired and trained by their employer.

Most windtechs complete certificates in wind energy technology, but some choose to earn an associate degree. Certificate programs last one year and associate degree programs in wind turbine maintenance are two years in length. Both programs provide classes that provide the skills needed to do the job of a windtech. The main difference is that general education courses are included in the associate degree program, but not in the certificate program.

There is no standard course of study, but typical coursework includes classes in the following:

- Basic turbine design
- Turbine diagnostics and repair techniques
- Control and monitoring systems
- Braking systems
- Electrical, mechanical, and hydraulic inspection and maintenance
- Computers and programmable logic control systems
- Rescue, safety, first aid, and CPR training

In addition to classroom studies, there is lab work. Some programs also give students hands-on training and practice on school-owned turbines and machinery. Other training may include an internship with a wind turbine servicing contractor.

Newly hired windtechs usually receive at least 12 months of on-the-job training related to the specific wind turbines they will maintain and service. Part or all of this training is provided by the manufacturer.

Professional certification is not mandatory, but it is preferred by some employers because it demonstrates a basic level of knowledge and competence. There are many organizations who offer certifications, in addition to the schools that include them in their training programs.

EARNINGS

INDUSTRY SOURCES REPORT THAT THERE is currently a shortage of trained windtechs, which has caused competition among many different employers. This has put experienced windtechs in a good position to command relatively high salaries. Currently, the median annual salary for windtechs overall is about $55,000, and the top 10 percent earn more than $80,000. Actual earnings depend on a number of factors, but are mainly affected by experience, education, and the size and type of company.

Experience is key to earning more. Industry sources report that entry-level positions pay beginning windtechs between $35,000 and $40,000 per year. Many windtechs start out as apprentices. Apprentice wages begin at 60 percent of full salary. Pay increases come along as apprentices learn to do more. With just a couple years of experience on the job, they usually break out of the beginner's pay range and get closer to the $50,000 a year level. Their salaries can continue to increase over time as their expertise and skills become more valuable to employers.

Some windtechs choose to increase their incomes and advance their careers by becoming engineers. That means going back to school to learn the necessary skills and earn a degree. This can represent a substantial investment in time and money, not to mention finding the time for school while continuing to work. The rewards are also substantial. Engineers in the wind power industry generally earn between $75,000 and $95,000 a year.

Median annual wages among the top industries that hire windtechs range from $50,000 to $60,000. The highest paying jobs are offered by electric power generation companies. The lowest are offered by service firms that handle repair and maintenance contracts. In between are utility system construction companies and various professional and technical services firms.

The majority of windtechs work full time and receive medical coverage and dental insurance. Some also receive bonuses, which can be as little as $800 a year or as much as $5,000. Profit sharing is uncommon, though it is sometimes offered. On average, profit sharing amounts to about $1,500 a year.

In addition to their regularly scheduled hours, windtechs are often on call to handle emergencies during evenings and weekends. Many are paid hourly wages rather than yearly salaries. For them, working outside normal hours means getting overtime pay that can nearly double their usual hourly rate.

Solar Installers

Overall, solar installers typically earn about $50,000 a year. Pay for inexperienced workers can start out as low as $35,000. After three or four years of working in the field, that rises to about $45,000. Solar installers with experience can earn more than $60,000. Depending on who you work for and whether pay is based on hourly wages, there might also be overtime pay.

Variations in wages are mostly dependent on geographic location and type of employer. For example, the highest salaries are received by solar installers in New York where the average pay rate is around $70,000 a year. The next five highest pay rates for experienced solar installers are in the states of Hawaii, New Jersey, Massachusetts, and Colorado. The lower paying states, like Vermont and Idaho, are in the $40,000 range.

Utility companies pay solar installers the most, typically offering salaries of about $65,000. Building construction is an industry that hires many solar installers, but most companies are relatively small. They generally offer $45,000 a year. At the low end are roofing contractors and product wholesalers, averaging $40,000.

OPPORTUNITIES

GREEN ENERGY IS A GOOD PLACE TO BE if you are looking for a career you can count on for the future. Solar panel installers and wind turbine technicians are the only occupations that are expected to double within the coming decade. The job growth in green energy is far outpacing other non-renewable occupations in the energy industry. In California alone, green energy employs more than half a million people. There are 10 times more green energy jobs in that one state than the number of coal mining jobs in the entire country. Nationwide, there are now more people employed in solar than in generating electricity through coal, gas, and oil combined. Clearly, green energy is the future.

The green energy industry has been boosted by big subsidies on wind and solar power, and general backing by the federal government since the turn of the 21st century. The Obama administration was particularly supportive of advancing solar power to create new jobs while reducing carbon pollution. While the Trump administration is planning to reverse or reduce this support, states are taking up the slack. Numerous states and local governments have passed supportive legislation to promote renewable resources. They are recognizing the benefits of solar and wind energy, including the creation of jobs. The greatest demand for workers can be found in states and localities that provide green energy businesses tax rebates, subsidies, and other incentives. If you are looking for the best place to take advantage of rising job rates, the most wind and solar friendly states are Hawaii, Idaho, Delaware, California, and Washington.

New jobs are being created as more and more solar panels are being installed. The cost of solar panels has been falling for a decade, and as that trend continues, more residential homeowners are taking advantage of the systems. There are even solar leasing plans that make owning a solar system much more affordable, creating additional demand for people to install them.

According to industry leaders, the job outlook for windtechs is very, very bright. The cost of wind power is coming down rapidly, making it much more competitive with coal, natural gas, and other forms of carbon-based power generation. There are now major wind farms in 41 states, and many more are under construction or in the planning stage. The American Wind Energy Association (AWEA) reports that the wind project construction and advanced development pipeline will be four times greater inside of one year.

As more wind turbines are erected, more windtechs will be needed to install and maintain them. Recent employment data show that the US wind sector is putting Americans to work at a breathtaking pace – a 32 percent increase in one year alone. Statistics predict that wind technician jobs will increase over 100 percent during the next decade, but as great as that is, some industry experts say the number will be higher.

The relatively low educational requirements for new windtechs make it an attractive occupation for anyone looking for a good job with a secure future. It also represents excellent opportunities for experienced energy workers to transfer their valuable skills to new jobs in green energy sectors like wind.

The building of new wind farms has been a particular boon to farming and ranching communities. While most occupations deliver the most jobs in urban areas, wind farms are generally located in rural areas and factory towns. There are definitely some areas that offer more opportunities than others. Windtechs need to go where the wind farms are, and they are generally more prevalent in the Great Plains, the Midwest, and along coasts. Currently, the best states are Iowa, Kansas, Oklahoma, and South Dakota, where wind is generating more than 30 percent of capacity. However, there are at least 14 states that are generating over 10 percent of electricity from wind, and wind's impact will surely be seen nationwide in the near future.

It is important to note that while entry into solar and wind occupations is easy, employers report difficulty hiring qualified workers. Candidates who complete courses at a community college or technical school will have the best job opportunities. Those with apprenticeships or experience as electricians, roofers, carpenters, or other construction work will also be the most competitive.

GETTING STARTED

GREEN ENERGY IS POISED FOR ENORMOUS growth, especially solar energy. In fact, solar photovoltaic is the fastest growing energy source in the world. This is clearly affecting the job market in a big way, but how can you take advantage of these new opportunities?

The most desirable solar installation jobs are in utility companies, but only the most experienced workers can expect to land those jobs. To join that elite group, it is necessary to start at the bottom in one of the many other solar-related industries. In general, solar power is especially fragmented, primarily made up of many very small companies like building contractors and solar panel distributors. The easiest way to find these companies is to search the big internet job sites. You will find literally thousands of job openings, many specifically asking for beginners. Considering no experience or education is usually required, their opening salary offers of $35,000 to $45,000 for entry-level installers is a good start.

The best way to impress a potential employer is to demonstrate passion for the green energy sector. Employers do not necessarily hire for skills, but rather look for an enthusiastic attitude. Most will teach candidates the necessary skills, but they do not want to waste their time on someone who just wants to make some money short term. The next best way to ace an interview is to get up to speed on the industry and the company. Do your homework and come prepared with well-informed questions. Learn the lingo. Even if you are passionate, if you cannot talk the talk, you probably will not get a job. You can pick up the common terminology in the industry by reading some solar energy blogs.

Do online research on photovoltaic power, inverters,

shading considerations, etc. The more knowledge you have, the more impressive you will be. Attend local conferences or classes on solar power. Many are advertised online. There are even free online courses, like those presented by Solar Energy International (SEI).

Solar installation employers prefer to see some experience. You can get that experience from volunteering in local community solar projects. If there are not any in your area, call a local solar energy company and ask to volunteer in some capacity. Your volunteer work may lead to a real job, but the goal is to learn as much as you can about the field and make contacts that can help you find a job.

To get your foot in the door quicker, get certified. Getting certified as an installer of solar energy components takes less time than going to college and you will start higher in the company from the start. Look online for technical programs and vocational schools near you that offer certifications. Some professional associations like The North American Board of Certified Energy Practitioners (NABCEP) offer online certifications for solar energy professionals.

Windtechs

Getting started in a wind turbine technician job is a little different because some training and experience are usually required. You can still find plenty of jobs through a simple online job search, but you should lay some groundwork to improve your chances before trying for an interview.

One of the best ways to start is by doing an internship with a wind company. An internship provides several benefits. First, you will have a chance to increase your knowledge and hone your skills. While getting some experience under your belt, you will also have a chance to show that you can be an asset. This may lead to a job offer, but you can also expect to get a glowing letter of reference and some good contacts to use in your job search. The AWEA website has a list of companies that offer internships on their Wind Energy Career Center page. There is also a list of current job openings.

Attending workshops is another good way to start. You can learn much through workshops, and they are also good for providing great networking opportunities. In fact, it is not unusual for an attendee to find a job as a direct result of attending a workshop. Even more potent networking opportunities can be found through national conferences. Conferences are extremely popular and often attract thousands of attendees. Some even offer first career workshops. Several hundred exhibitors may be at a conference – each one represents an opportunity to meet a potential employer. Introduce yourself to as many as possible. You will be surprised by how many say they are hiring.

ASSOCIATIONS

- **Solar Energy International (SEI)**
  https://www.solarenergy.org

- **The North American Board of Certified Energy Practitioners (NABCEP)**
  www.nabcep.org

- **American Solar Energy Society (ASES)**
  https://www.ases.org

- **American Wind Energy Association (AWEA)**
  www.awea.org

- **Interstate Renewable Energy Council**
  https://irecusa.org

PERIODICALS

- **Solar Today**
www.ases.org/solartoday

- **Renewable Energy World**
www.renewableenergyworld.com

WEBSITES

- **Solar Ready Vets**
https://www.energy.gov/eere/solar/solar-ready-vets

- **Wind Exchange**
https://windexchange.energy.gov

- **Renewable Energy Jobs**
www.energyplacement.com

- **Dayaway Careers**
https://www.dayawaycareers.com

Copyright 2019
Institute For Career Research
CAREERS INTERNET DATABASE

www.careers-internet.org

www.ingramcontent.com/pod-product-compliance
Lightning Source LLC
Chambersburg PA
CBHW071201220526
45468CB00003B/1119